Where Are the Constellations?

by Stephanie Sabol

illustrated by Laurie A. Conley

Penguin Workshop

For Adam and Maya, my stars—SS

For Brian J. and Philip—LAC

PENGUIN WORKSHOP
An Imprint of Penguin Random House LLC, New York

Published by Penguin Workshop, an imprint of Penguin Random House LLC, New York.
PENGUIN and PENGUIN WORKSHOP are trademarks of Penguin Books Ltd.
WHO HQ & Design is a registered trademark of Penguin Random House LLC.
Printed in the USA.

Visit us online at www.penguinrandomhouse.com.

Library of Congress Control Number: 2021005878

ISBN 9780593223734 (paperback) 10 9 8 7 6 5 4 3 2 1
ISBN 9780593223741 (library binding) 10 9 8 7 6 5 4 3 2 1

Contents

Where Are the Constellations?

Thousands of years ago, there were no calendars or clocks as we know them. People couldn't tell time by glancing at their cell phones. Instead, they looked for clues in the sky—most often by looking at stars—to tell them what time of day or even what time of year it was. The ancient Egyptians, for example, eagerly waited for a certain star to appear. Once it did, it meant the flooding of the Nile River was about to start. When the flood receded, the soil was fertile and ready for planting crops.

There were also no accurate maps. However, the ancient Greeks noticed that one star always appeared in the same place—we call it the North Star, or Polaris (Polar Star). Sailors used it as a guide to help them figure out in which direction they were heading.

Today, in cities, lights and pollution blot out many stars. However, on a clear night a person who is in the country can see thousands. Still, finding one star among thousands can be hard. So long ago in different cultures people looked for groups of stars, and they imagined them forming pictures by drawing lines from one star to another. It was like a connect-the-dots puzzle today. These "pictures" were easier to spot than just one star. Each picture was given a name and told a story.

Different cultures had different names and stories for the same group of stars. To people in North America, a group of seven stars looked like a long-handled soup spoon. They called it the Big Dipper (*dipper* is a word for this kind of soup spoon). However, in England, the same stars were known as the Great Plough. (A plough, more commonly spelled *plow* in the United States, is a farming tool that breaks up soil.) In Germany, the stars were known as the Great Cart.

Big Dipper

Star pictures are known as constellations. The scientific study of stars—and everything else in the sky—is called astronomy. Astronomers of the twenty-first century know far more about the stars than ancient peoples did. They have learned what stars are made of, how long they live, and how they die. Yet despite all that has been discovered, even astronomers—just like everyone else—still look to the night sky with the same sense of wonder.

CHAPTER 1
Our Address in Space

Earth's atmosphere

Space. What is it? Well, it's everything in the sky that is above Earth's atmosphere—a layer of gases that is about sixty miles thick. Nobody knows exactly how big space is. But it's enormous, really too big for our minds to grasp. So, it's helpful to think of Earth as having an address in space, just like when a person has an address where they live.

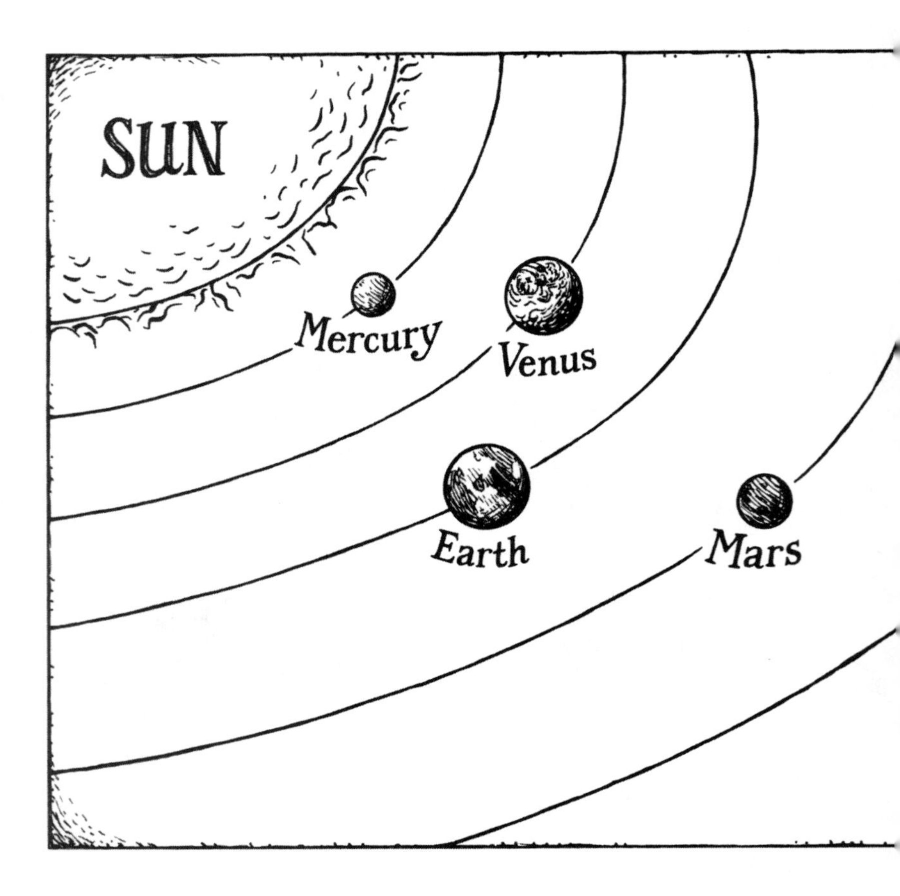

Earth—the planet we live on—is part of our solar system. Our solar system contains eight planets that orbit (circle) just one star, the sun. A star is a ball of hot gas—mostly hydrogen and helium. Even though the sun seems huge to us,

Our solar system

it's actually a medium-size star. And since it's 93 million miles away from us, it's the closest star to Earth. During the day, the light from the sun is so bright, it blocks out other stars.

But at night, after the sun has set, many other stars in the sky can be seen.

Our solar system is part of a galaxy, a city of stars. There are many galaxies. Our galaxy is known as the Milky Way. It was named because of the whitish hue it gives off that looks like milk spilled across the sky. It contains at least 100 billion stars!

The Milky Way looks like a pinwheel with four long arms. Our solar system is located in one of the four arms. It orbits around the center of the Milky Way at about 515,000 miles per hour. Even at that very fast speed it will take our solar system 230 million years to orbit all the way around the Milky Way's center. Dinosaurs roamed our planet the last time our solar system was in this spot!

What is bigger than a galaxy? A group of galaxies. The Milky Way is part of a cluster of more than thirty galaxies called the Local Group.

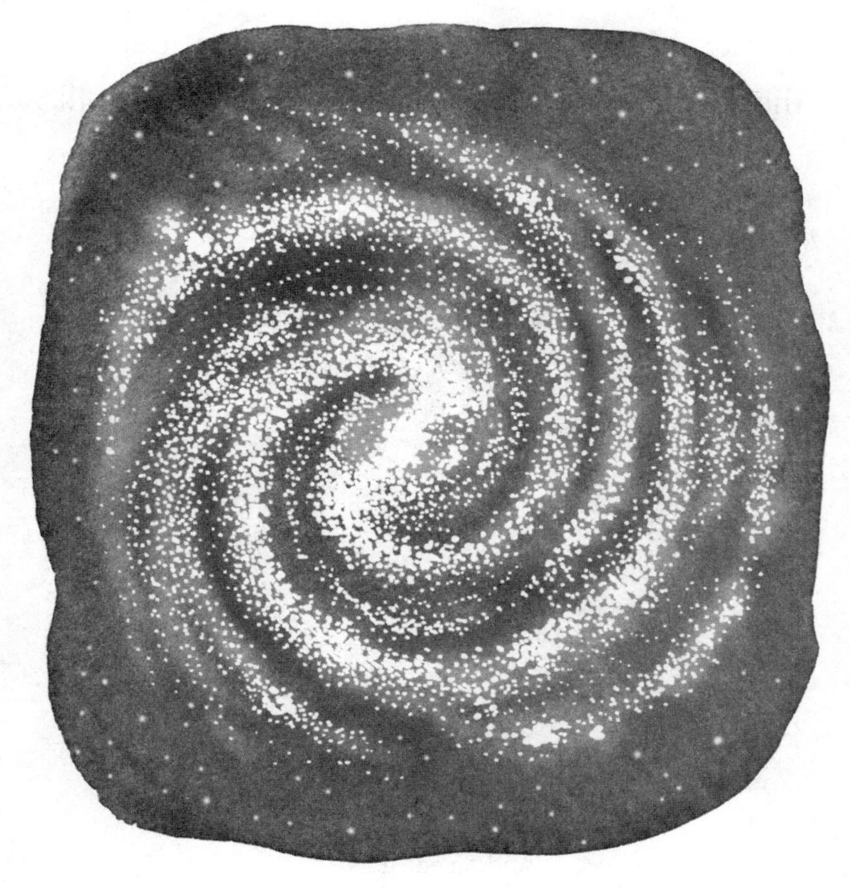

Milky Way

This group contains the Andromeda galaxy. Of big galaxies, it is the one closest to the Milky Way. (The Andromeda galaxy is twice as wide as the Milky Way.) Scientists predict that in the next few billion years the two galaxies will collide and form a super-galaxy.

Andromeda galaxy

As big as the Local Group is, it's just a small part of a larger group of galaxy clusters called the Virgo Supercluster. It's an enormous collection of more than two thousand galaxies.

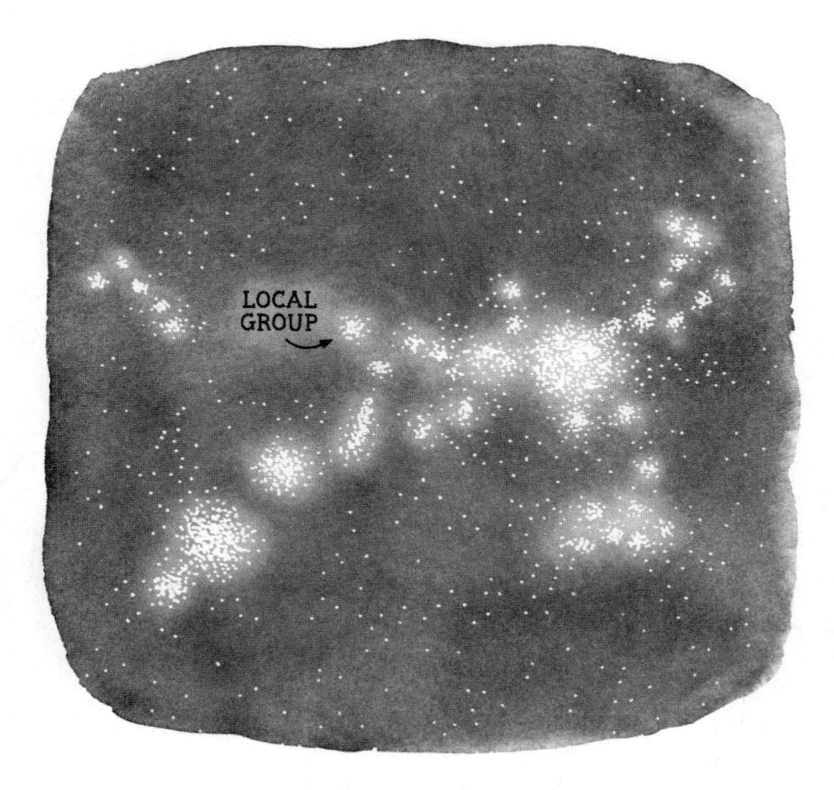

Virgo Supercluster

These galaxies are all part of our universe. The universe includes everything—Earth, our solar system, the Milky Way galaxy, and beyond. The universe is everything in space that we can and can't see.

The Story of Andromeda

Andromeda was a princess in an ancient Greek myth. She was the daughter of Queen Cassiopeia. Queen Cassiopeia believed that her daughter was more beautiful than the daughters of the sea god, Poseidon. That made the sea god so angry, he wanted to destroy the queen's land. What was Cassiopeia to do? She asked an oracle, a person who was able to communicate directly with the gods. The oracle said the only way to gain the sea god's

forgiveness was to sacrifice her daughter. So the queen chained Andromeda to a rock by the ocean where a sea monster named Cetus would come and kill her. Luckily, however, a brave young man named Perseus was passing by. He offered to slay the sea monster in return for Andromeda's hand in marriage. Perseus defeated the sea monster and did indeed marry Andromeda.

How did the universe start? No one knows for sure, but astronomers have an idea, or theory. It's called the big bang theory. This theory originally stated that 13.8 billion years ago, the entire universe was packed into a tiny point that was much smaller than the period at the end of this sentence. This point contained tiny particles that were mixed with light and energy. The particles were extremely hot and packed together. In just an instant they began to expand.

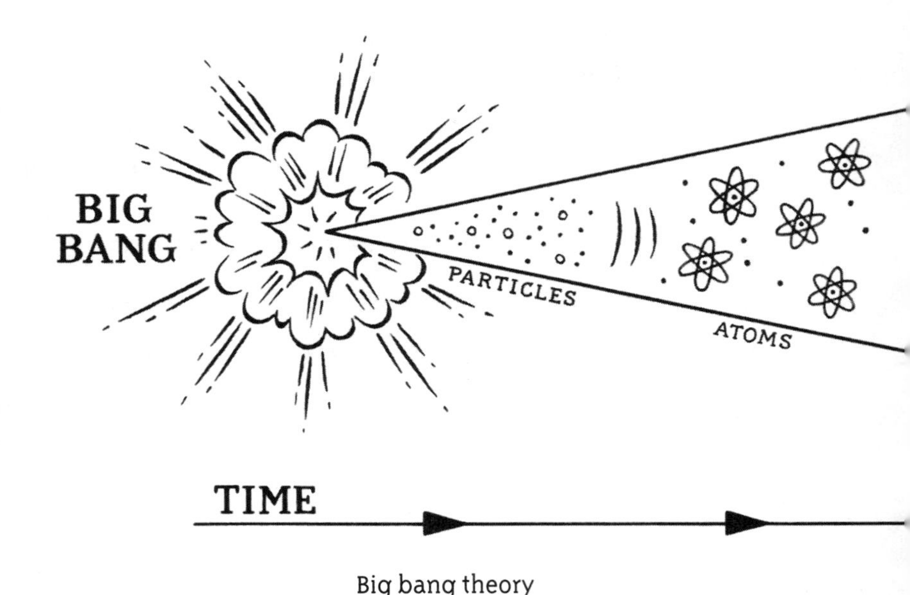

Big bang theory

This expansion is known as the big bang. It was the beginning of space and time. The particles would eventually become what we know as the planets and stars and everything else. The universe still continues to expand today, just like a balloon does when it is blown up. But scientists continue to update this theory as they observe and collect more data.

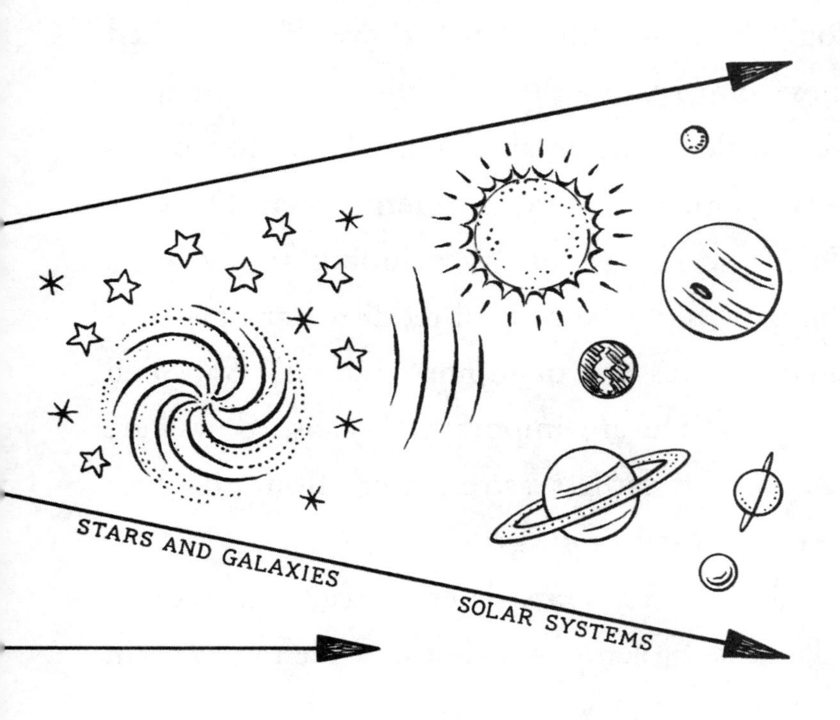

CHAPTER 2
Star Tracking

One of the early records of star tracking began in 350 BC. A Chinese astronomer named Shi Shen noted about eight hundred stars that could be seen with the naked eye. He compiled these into one of the first known star catalogs. Very little is known about Shi Shen. He worked with another Chinese astronomer, Gan De. Gan De is said to have observed Jupiter and what he thought might be a small reddish star next to it (likely it was one of Jupiter's moons). Shi Shen and Gan De are important because they were two of the earliest astronomers from the East to record their findings.

About 450 years later, Greek astronomer Claudius Ptolemy was born. Ptolemy lived in

Egypt when it was part of the Roman Empire. Around the year AD 150, he wrote the *Almagest*. In the *Almagest*, Ptolemy identified more than a thousand stars that made up forty-eight constellations. He chose names for the constellations based on Greek mythology as well as animals and objects that were commonly known to the ancient Greeks.

Claudius Ptolemy

Pegasus

For example, he named one constellation Pegasus after a legendary white winged horse. Pegasus struck his foot against a rock and created a magical fountain. All who drank from this fountain were blessed.

Naming the constellations wasn't what Ptolemy was best known for. He studied the movement of other planets. From what he observed, Ptolemy thought those planets as well as the sun rotated around Earth. He thought Earth was the center of the solar system. Ptolemy was brilliant, but wrong! It would be another 1,400 years before anyone realized it.

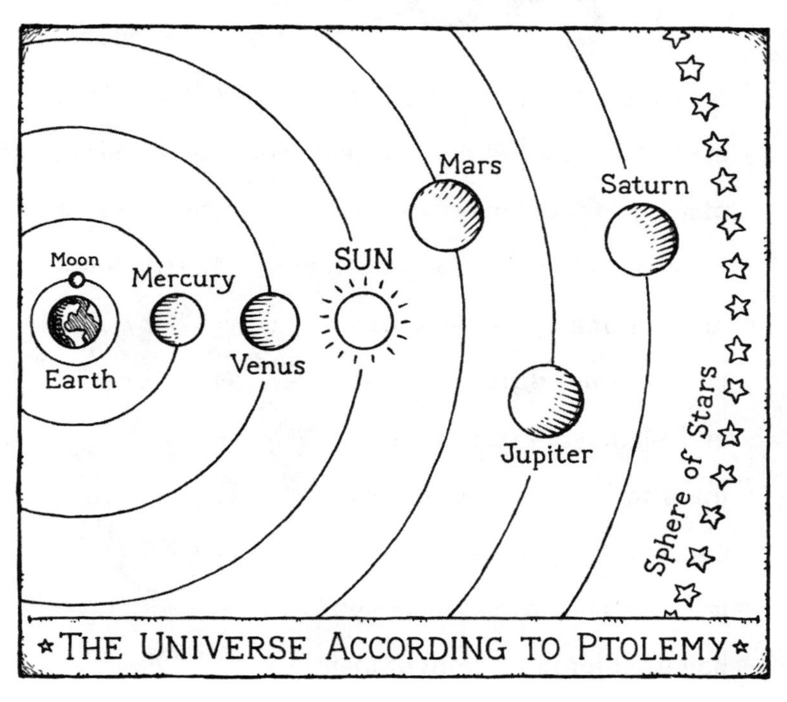

THE UNIVERSE ACCORDING TO PTOLEMY

Star Trackers Unite!

More than a hundred years ago, in 1919, the International Astronomical Union (IAU) was formed.

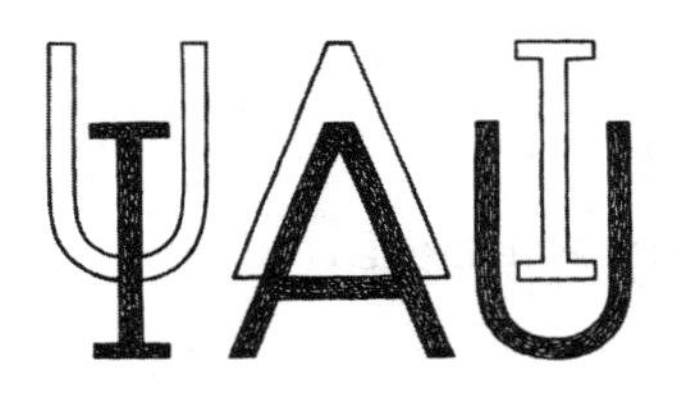

The IAU was created by several countries to promote the science of astronomy. It is a highly respected group. In 2006, it demoted Pluto. That made big news. Pluto had been listed as the ninth planet in the solar system till then. But the IAU said it wasn't actually a real planet, only a dwarf planet.

Another one of the IAU's jobs is to name different stars and planets. In 1922, it came up with eighty-eight official constellation names. The IAU used the same names Ptolemy had used for forty-eight of them.

Pluto

In 1543, a Polish astronomer named Nicolaus Copernicus published a book that disagreed with Ptolemy. Copernicus understood that the sun was a star. He said that the sun, not Earth, was the center of the solar system. He pointed out that Earth moved in an orbit around the sun,

Nicolaus Copernicus

just as the other planets do. Copernicus's book was published just two months before he died.

Copernicus didn't live to see the reaction to his work, but it might have been better that way. People were furious that he suggested the sun, not Earth, was the center of the universe. How could Earth not be the center? That's where all the people were!

Galileo Galilei

Decades later, another astronomer helped prove that Copernicus was right. His name was Galileo Galilei. He was born in Italy in 1564. Galileo heard about a man from the Netherlands named Hans Lippershey who made eyeglasses. Lippershey took two of his eyeglass lenses and put them on either side of a tube. Looking inside this tube made everything look bigger! This was the first telescope. Galileo improved on Lippershey's telescope so that it could magnify objects to twenty times their size!

Hans Lippershey

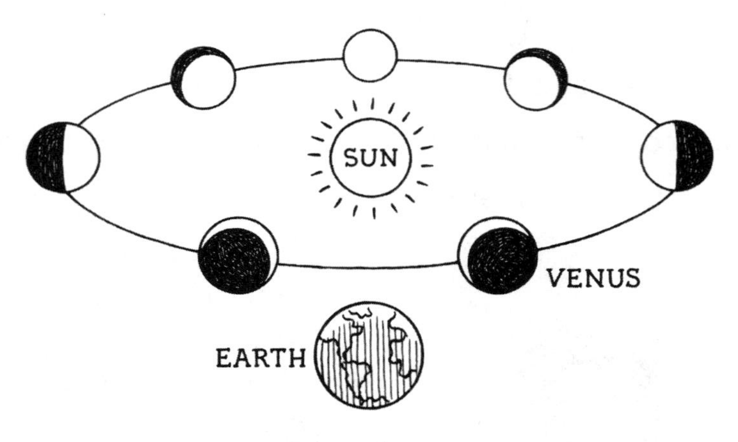

Phases of Venus

Using his telescope, Galileo noticed that the planet Venus changed shape depending on what time of the month it was. For Galileo, this proved that Venus was orbiting the sun and reflecting the sun's light. That was the cause of what appeared to be changes in the planet's shape. Galileo helped support Copernicus's theory that the sun was indeed the center of our solar system. The Catholic Church was unwilling to believe that Earth was not the center of the universe. Galileo was put under house arrest for the rest of his life!

Galileo on trial

As scientific tools became more advanced, astronomers and explorers were able to locate more stars. In 1603, German lawyer and astronomer Johann Bayer created a beautiful star atlas called *Uranometria*. The atlas included illustrations of each of Ptolemy's forty-eight constellations. In addition to these, *Uranometria*

also included twelve other constellations. They could only be seen from the Southern Hemisphere. They'd been discovered by Dutch explorers in the late 1500s as they traveled below the equator to the East Indies. They were named after exotic creatures like a peacock and a chameleon.

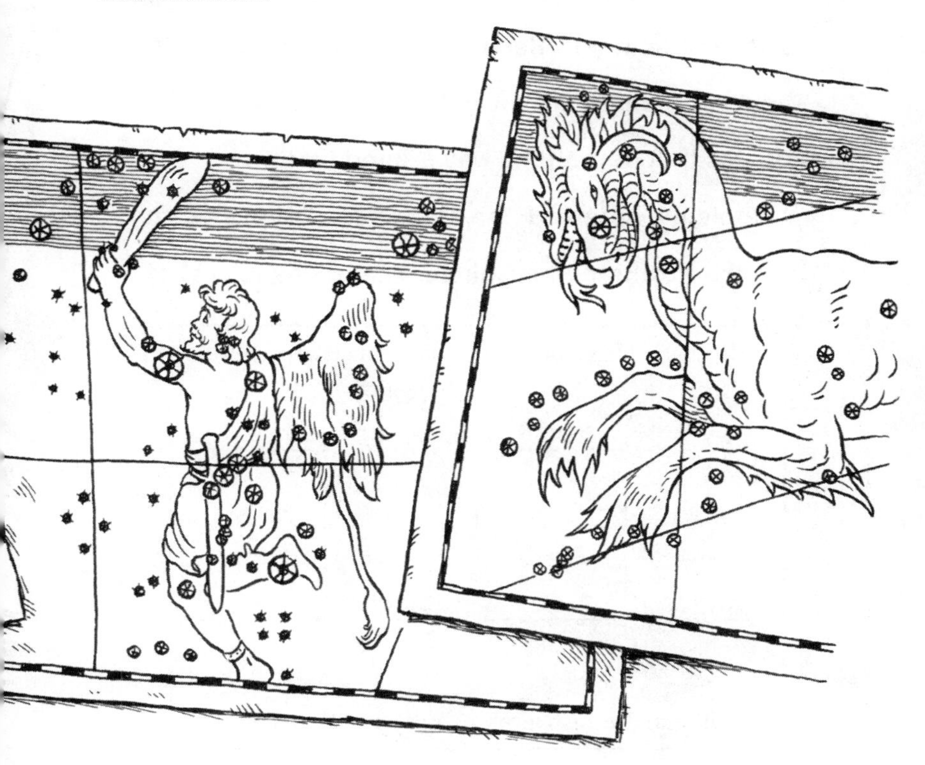

Pages from Johann Bayer's book *Uranometria*

Hemispheres

Not everyone on Earth sees the same stars and constellations. It depends where on the planet you live. Think of Earth as being sliced through the middle. The Northern Hemisphere is the top half of Earth. The Southern Hemisphere is the bottom half. They are divided by an imaginary line called the equator that runs all the way around Earth.

Ptolemy observed the stars from Egypt, which is in the Northern Hemisphere. That's why his list of

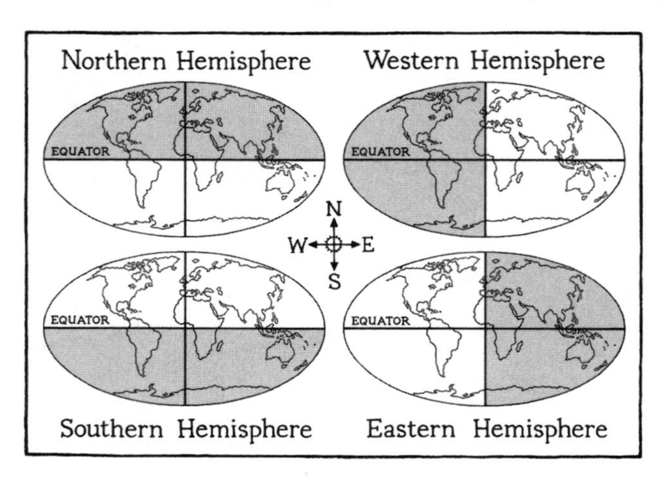

constellations doesn't include any that can only be seen from the Southern Hemisphere.

Nicolas Lacaille was a French astronomer who observed the sky from southern Africa in the 1750s. He named another fourteen constellations that are found in the Southern Hemisphere and are not very bright. Lacaille named his constellations after scientific tools like a clock, an air pump, and an octant, which is a navigation tool for sailors.

If you located a new constellation, what would you name it?

Nicolas Lacaille

CHAPTER 3
Our Star, the Sun

Even though we don't see the sun in the night sky, the sun *is* a star. Our sun was born 4.6 billion years ago—nearly 10 billion years after the big bang. Scientists believe that a star forms when a cloud of dust and gas begins to spin quickly. The center turns into a star.

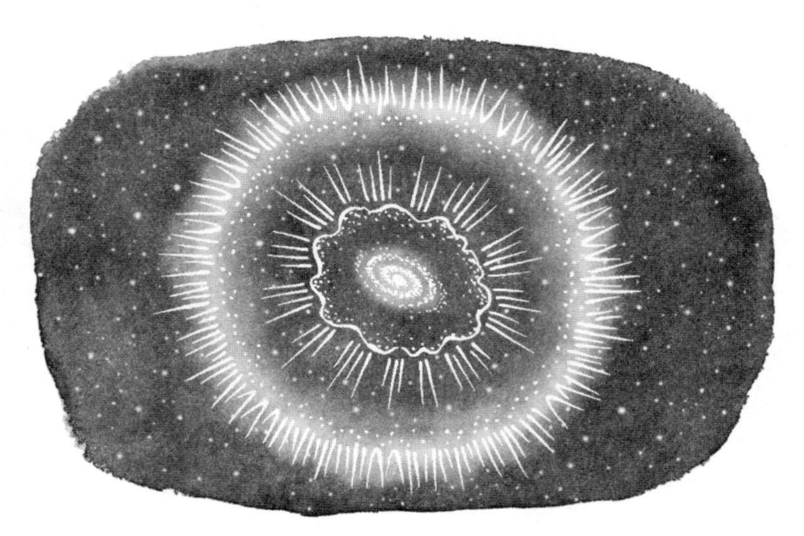

The formation of a star

Like all stars, the sun is a huge ball of very hot gas. The temperature at its core is around 27 million degrees Fahrenheit. Even though the outside is much cooler (10,000 degrees), that's still way hotter than human beings can begin to imagine! Many people find it uncomfortable when the temperature reaches the nineties.

The sun is nearly 870,000 miles across. If the sun were the size of a basketball, Earth would only be the size of the head of a pin!

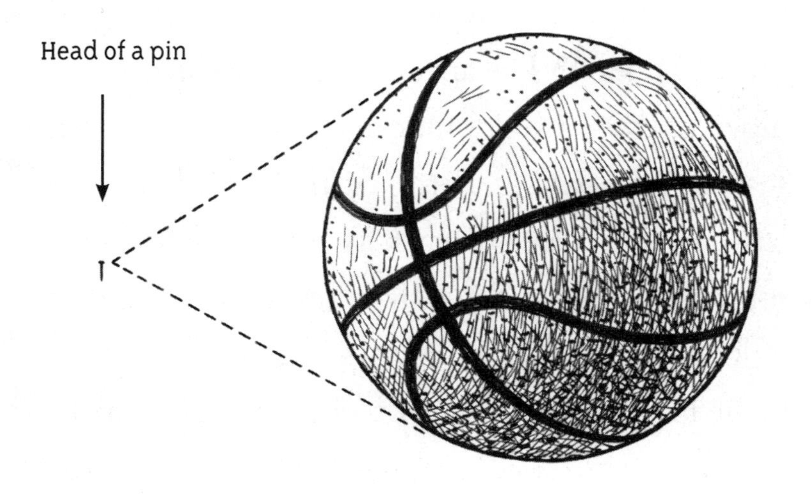

And even though it's made of gas, the sun is 330,000 times heavier than Earth.

Most people would say the sun looks yellow. (Remember, no one should ever look at the sun directly.) But the sun is actually white. Light is a mixture of the seven colors of the rainbow. However, the air surrounding Earth filters out some of the colors. The result is a sun that looks yellow to us.

Among the most beautiful sights on earth are sunrises and sunsets. Just as it appeared to ancient people, to us the sun seems to move across the sky. But does it? The answer is no. It's *Earth* that is moving. Every twenty-four hours, Earth completes a spin. The part of Earth facing toward the sun has day. The part of Earth facing away from the sun has night. That is why when it is the middle of the day in New York, it's the middle of the night in Tokyo, which is on the other side of the planet.

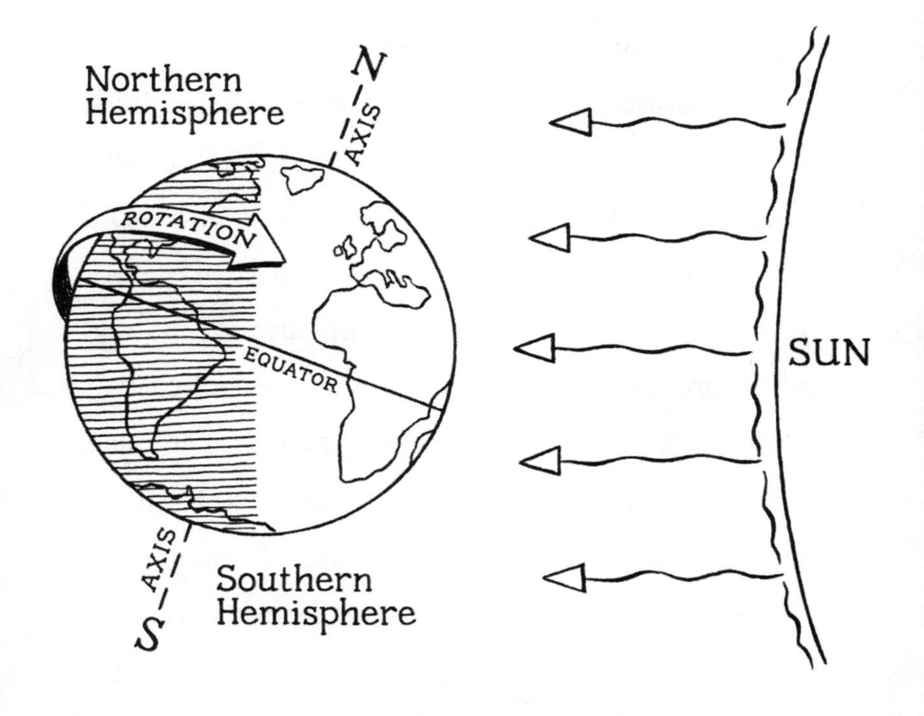

Also revolving around the sun are the seven other planets in our solar system. Unlike stars, planets can't make their own light. The planets in our solar system were probably made from the leftovers of the same gas and dust that formed the sun.

The planet closest to the sun is Mercury. After that come Venus, Earth, and Mars. These four are known as the inner planets. All are made of hard material—rock and metal. That is why Earth is called the "third rock from the sun." The other four planets are the outer planets—Jupiter, Saturn, Uranus, and Neptune. They have solid cores but are mostly made of gas. The outer planets are much bigger than the inner planets. Our solar system also has more than two hundred moons! Most of these orbit around one of the planets. Moons come in many shapes and sizes.

Earth's moon helps control Earth's weather—a very important job that enables us to have life on our planet.

Are there other solar systems? Most definitely. Ours is the one officially called "solar system" by astronomers. But they have discovered more than 2,500 other stars with planets orbiting them in the Milky Way galaxy.

Of course, to people on Earth, the sun seems like it's the center of the universe. To us, our star is truly the star of the show!

The moon orbiting Earth

Fun Facts about the Planets

Mercury: Because it's closest to the sun, Mercury takes just eighty-eight Earth days to circle the sun.

Venus: Venus is often mistaken for a star. After the sun and moon, it is the brightest thing people can see in the sky.

Earth: Earth is exactly the right distance from the sun (not too close, not too far away) so that life can thrive. Other planets are either too hot or too cold to allow for this to happen. It has only one moon.

Mars, the god of war

Mars: Mars is referred to as the Red Planet. The soil on Mars contains iron, which looks red when it rusts. Because red is the color of blood, the ancient Romans named the planet Mars after their god of war.

Jupiter: Jupiter is the first of the outer planets. Jupiter is so massive that all the other planets could fit inside it. Jupiter has seventy-nine known moons, more than any other planet in the solar system.

Saturn: Saturn is one of the most beautiful planets due to its rings. These rings are made up of ice, dust, and pieces of rock.

Saturn

Uranus: Uranus orbits in an unusual way. It is tilted so that its southern hemisphere is always facing the sun. Scientists think that long ago something big must have crashed into it and knocked Uranus over.

Neptune: Neptune is the farthest planet from the sun—it takes nearly 165 Earth years for it to complete one orbit!

CHAPTER 4
The Life of a Star

The sun is our special star, but there are trillions more stars in the universe. How does a star begin its life?

Over millions of years, gravity squeezes dust and gas particles together into clouds. Gravity is the invisible force that pulls objects together. Without gravity, everything (including human beings) would float away! Clouds in space formed by gravity are called nebulas. Sometimes nebulas are called "star nurseries." A nebula begins to heat up as it revolves. This is a baby star, or protostar.

Once the protostar gets hot enough, it releases energy—what we call starlight. From Earth, stars may look like they are twinkling—

Swan Nebula

but that's just an illusion. Out in space, stars shine steadily. As their light travels through Earth's atmosphere, it gets bent by layers of hot and cold air. That's what makes them look like they are glittering.

Stars grow to different sizes. The size of the star is what determines how long the star will live and also how it will die. Our sun—remember, it is a medium-size star—will likely live for another

Protostar

five billion years. Its gas will burn up, but it will continue to grow to more than 100 times bigger. It will then be a red giant. That will be the sun's grand finale. Once the red giant runs out of fuel, it will shrink down to a fraction of its original size and become a white dwarf.

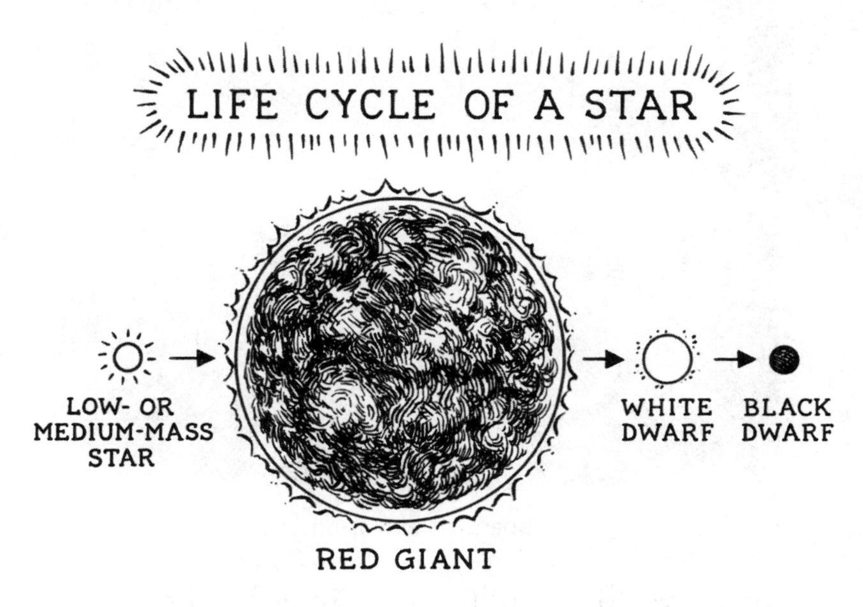

Then, a few billion years later, the sun will be nothing but a black dwarf—a cold, dark cinder that gives off no light.

What happens to stars larger than our sun? Some have deaths that are more dramatic than that of a medium-size star. They become supernovas and explode in a huge flash of light.

Supernova explosion

Supernovas burn millions of times brighter than average stars. Supernovas are rare, so astronomers get very excited by them. You don't have to be an astronomer to spot one. In 2008, a teenager from New York spotted a supernova using a small telescope. She was the youngest person to find one!

Caroline Moore, the youngest person to find a supernova

As for the biggest stars, some turn into a black hole when they die. They get smaller and denser until they become tinier than the head of a pin!

That pinhead has very strong gravity and sucks *everything* (even light) around it inside, just like a drain sucks water. If another star gets too close to a black hole, it will get sucked into it, too. Do people on Earth have to worry about getting sucked into a black hole? No. Scientists don't think there are any black holes close enough to Earth for that to happen.

What are some stars people see at night? The brightest star we can see is Sirius, or the Dog Star. It's called the Dog Star because it's part of

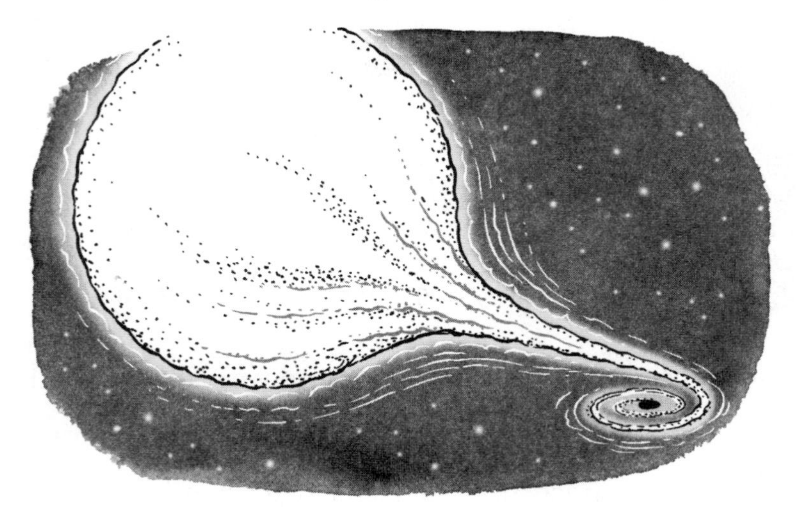

Star being destroyed by a black hole

a larger constellation of stars that looks like a dog! At 8.6 light-years away, Sirius is one of the sun's closest neighbors. Sirius is only twice as large as the sun, but gives out more than twenty times as much light.

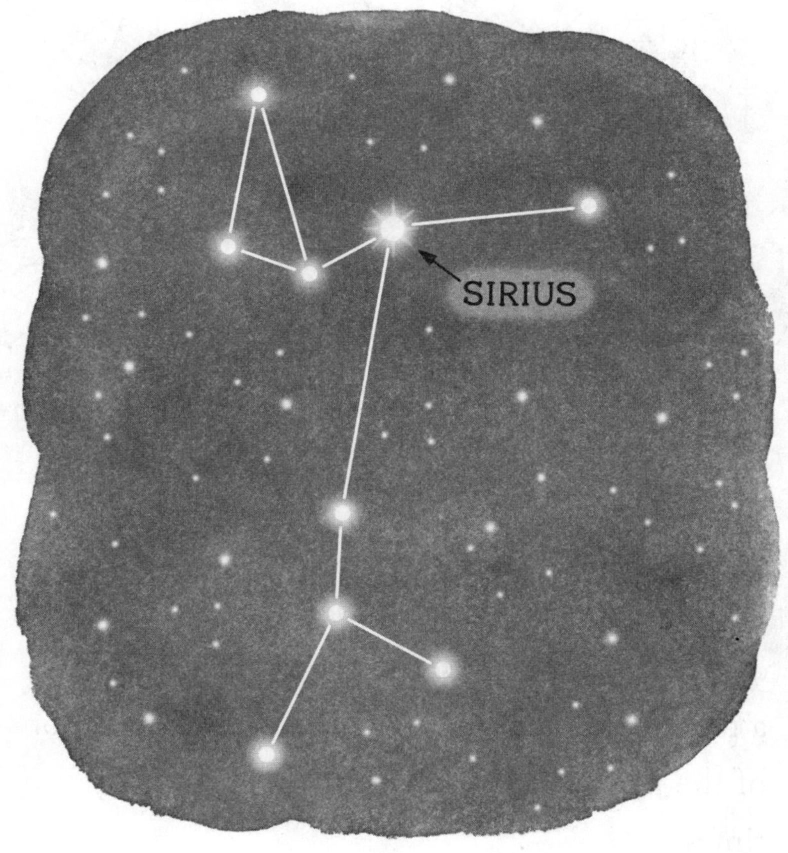

Sirius, the Dog Star

In legends from India, the Dog Star is known as Svana. The gods sent the dog Svana to serve a prince. Svana remained the faithful companion of the prince for many years, through many bad times.

When one of the gods came to take the prince to heaven, the prince was told there was no room for the dog. He replied that if there was no room for the dog, there was no room for him. It turned out that this was all a test, made up by the gods. And the prince's answer was the right one. He passed the test—and was allowed into heaven. As for Svana, the dog was honored with a place among the stars.

Proxima Centauri is the closest star to the sun. But it's still very far away. The light from Proxima Centauri takes 4.3 light-years to reach the sun. By comparison, the sun's light only takes a little over eight light-minutes to reach Earth.

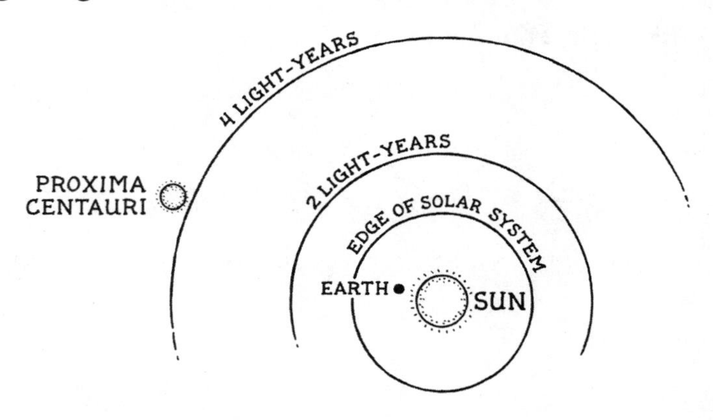

Light-Years

To calculate the vast distances between objects in space, astronomers needed a much bigger unit of measurement than miles. They came up with the concept of light-years. A light-year is the distance light travels in one Earth year. That's almost six trillion miles.

The idea of light-years can be tricky to understand. If a star is fifty light-years away, it means that the light from the star won't reach Earth for fifty years. When it does, we will be seeing what the star looked like fifty years ago, not what it looks like right at that moment!

Betelgeuse is a star located about seven hundred light-years away. It's ten million years old, but that is considered a pretty young star. Lately, it has been acting strangely. It used to be one of the top ten brightest stars, but not anymore. It's not even in the top twenty. Why is it fading? At first scientists thought Betelgeuse was dying. Now they believe its light was temporarily blocked by a cloud of hot gas. One day Betelgeuse will die. It might not happen for one hundred thousand years. In the lifetime of a star, that's just an instant!

Betelgeuse

Sirius, Proxima Centauri, and Betelgeuse are important stars on their own. All three are also parts of constellations.

Meteors

Sometimes you might see what looks like a star falling toward Earth. But it's actually not a star at all. It's a meteor. A meteor is a leftover piece of debris from the solar system. This debris might be from a comet (a chunk of ice that orbits the sun) or broken asteroid (a piece of rock that orbits the sun). As the debris falls through the sky it reaches Earth's atmosphere and flares up into a bright burst of light. Most meteors burn up in the atmosphere, but some are hard enough to fall to the ground. Then they are known as meteorites.

CHAPTER 5
Constellation Families

The British author John Berger wrote that the ancient people "who first invented and then named the constellations were storytellers. Tracing an imaginary line between a cluster of stars gave them an image and an identity. The stars threaded on that line were like events threaded on a narrative." In other words, each constellation told a story.

John Berger

Many of the constellation stories come from the myths of ancient Greece and Rome, and the Arab world. They knew many of the same stories as other ancient people did, like the Sumerians and Babylonians.

In 1975, an American astronomer named Donald Menzel wrote a book titled *A Field Guide to the Stars and Planets*. It grouped the eighty-eight constellations that the IAU described into eight constellation families. It was an easy way to help stargazers—whether in the Northern or Southern Hemisphere—locate different constellations.

Donald Menzel

Ursa Major is the only constellation family located entirely in the Northern Hemisphere. The Ursa Major family contains ten constellations, one of which is also called Ursa Major. *Ursa Major* means *Great Bear* in Latin.

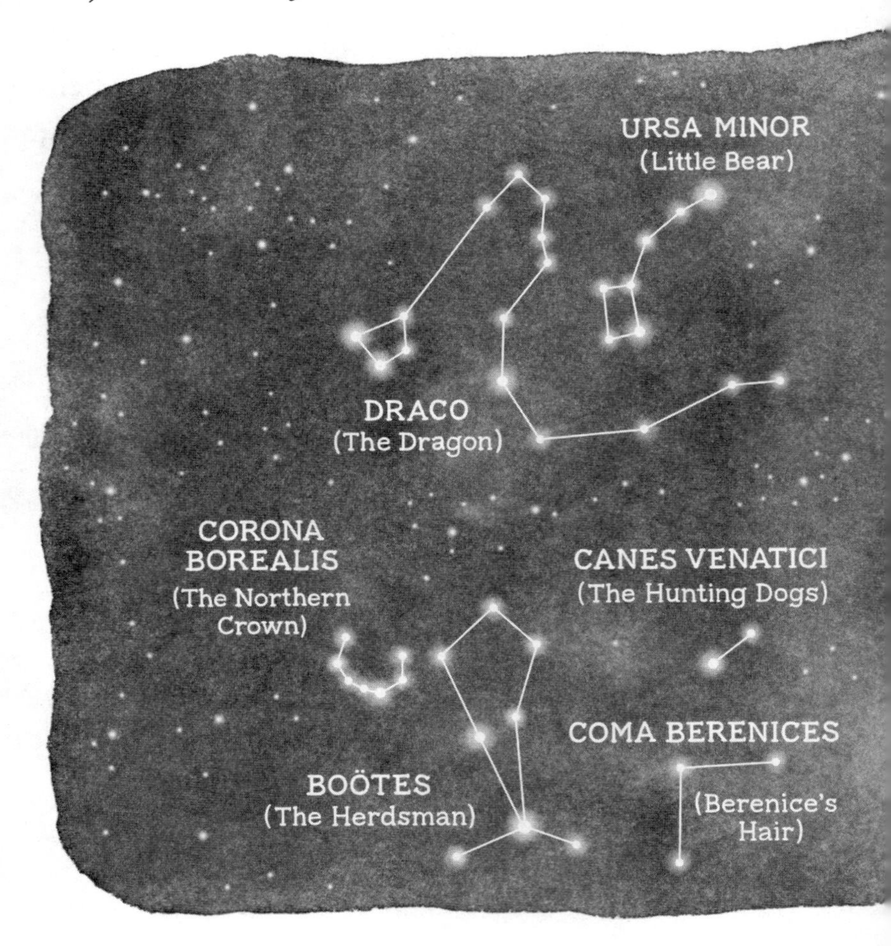

Ursa Minor means *Little Bear*. The Big Dipper is actually part of Ursa Major. The Big Dipper is called an asterism—that's the name for a smaller group of stars located within a constellation.

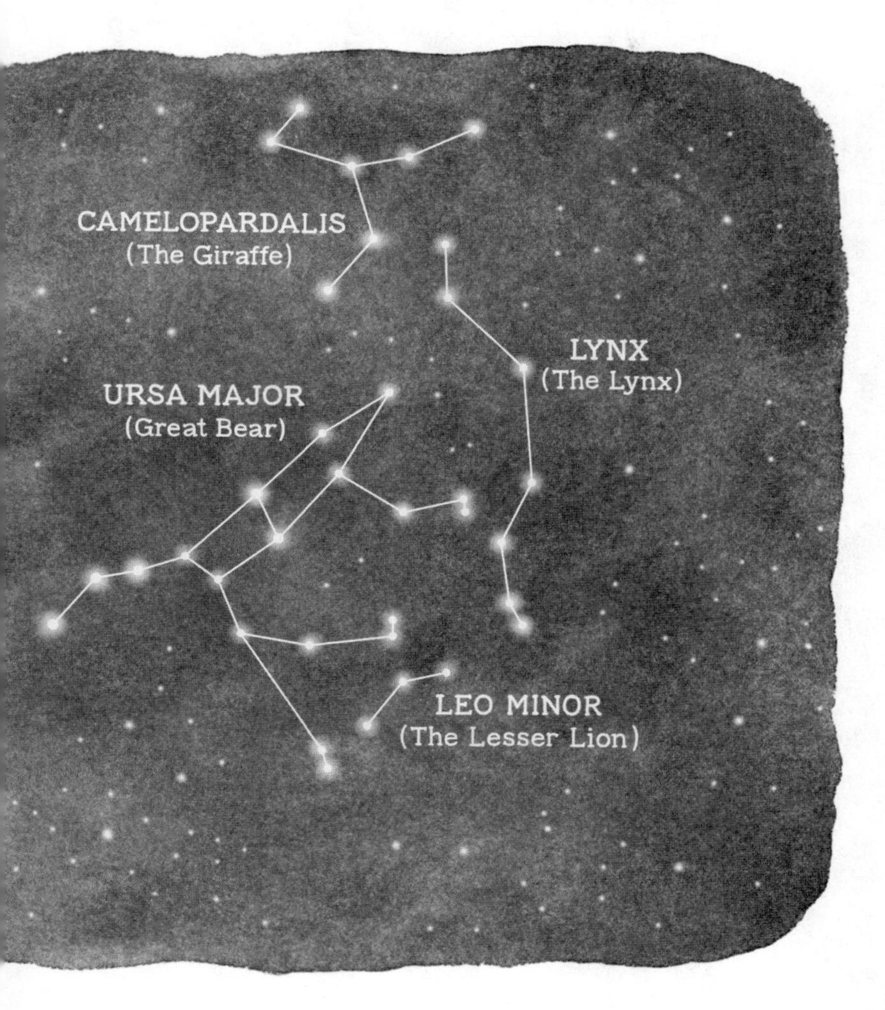

Ursa Major is the third biggest constellation in the sky. It is highest in the sky in the summer and lowest in the autumn. A Native American legend says this is because the bear is looking for a place to hibernate. According to Roman legend, a jealous goddess turned a beautiful maiden into a bear; later on, her son was also turned into a bear and both were placed in the sky.

Besides Ursa Major there is Ursa Minor, the smaller sibling of Ursa Major. It looks like a miniature version of the Big Dipper. Ursa Minor includes the North Star, Polaris.

The constellation family Perseus features many stories from Greek mythology. Of its nine constellations, only one, Cetus, is located in the Southern Hemisphere. The others are located in the Northern Hemisphere. Remember that Perseus was the Greek hero who saved Andromeda from the sea monster, Cetus. Both Andromeda and Cetus are constellations located in this family.

Cetus: The Sea Monster

Not only did Perseus kill a sea monster, he also managed to slay a fearsome monster called Medusa, whose hair was made of snakes. Anyone who looked straight at Medusa turned to stone.

So how did Perseus kill her? When he found Medusa asleep, he used his shield as a mirror, which he positioned so that he could see her reflection. Then, with a backward thrust of his sword, he finished her off. Pretty clever!

The constellation family Perseus

LACERTA
(The Lizard)
ANDROMEDA
(The Chained Maiden)
PEGASUS
(The Winged Horse)
TRIANGULUM
(The Triangle)
CETUS
(The Sea Monster)

With nineteen constellations in it, the Hercules constellation family is the biggest. It is located equally in both the Northern and Southern Hemispheres. Hercules is one of the most well-known heroes in classical mythology. (The Greeks called him Heracles, but we know him better by the Roman version of his name, Hercules.) Half god and half human, Hercules was the son of Zeus, king of the gods. Hercules was known for his strength. He was told to do twelve almost impossible tasks called labors. These labors included capturing and killing many monsters. Hercules succeeded at them all.

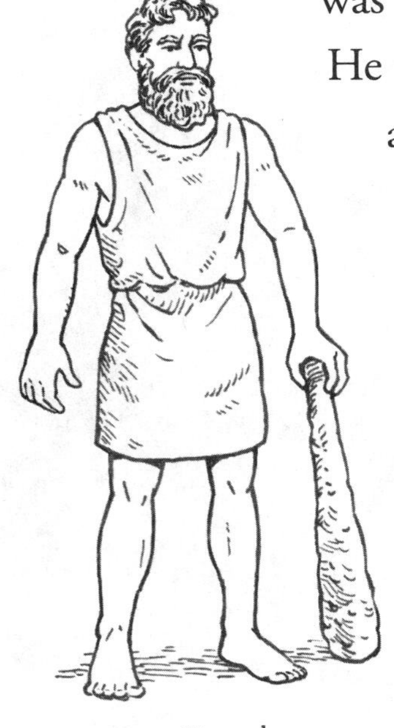
Hercules

Hercules is both the name of the larger family as well

as a constellation within the family. The stars that outline Hercules's constellation show him kneeling over a dragon he killed. Hercules is a tricky constellation to find since many of its stars are faint.

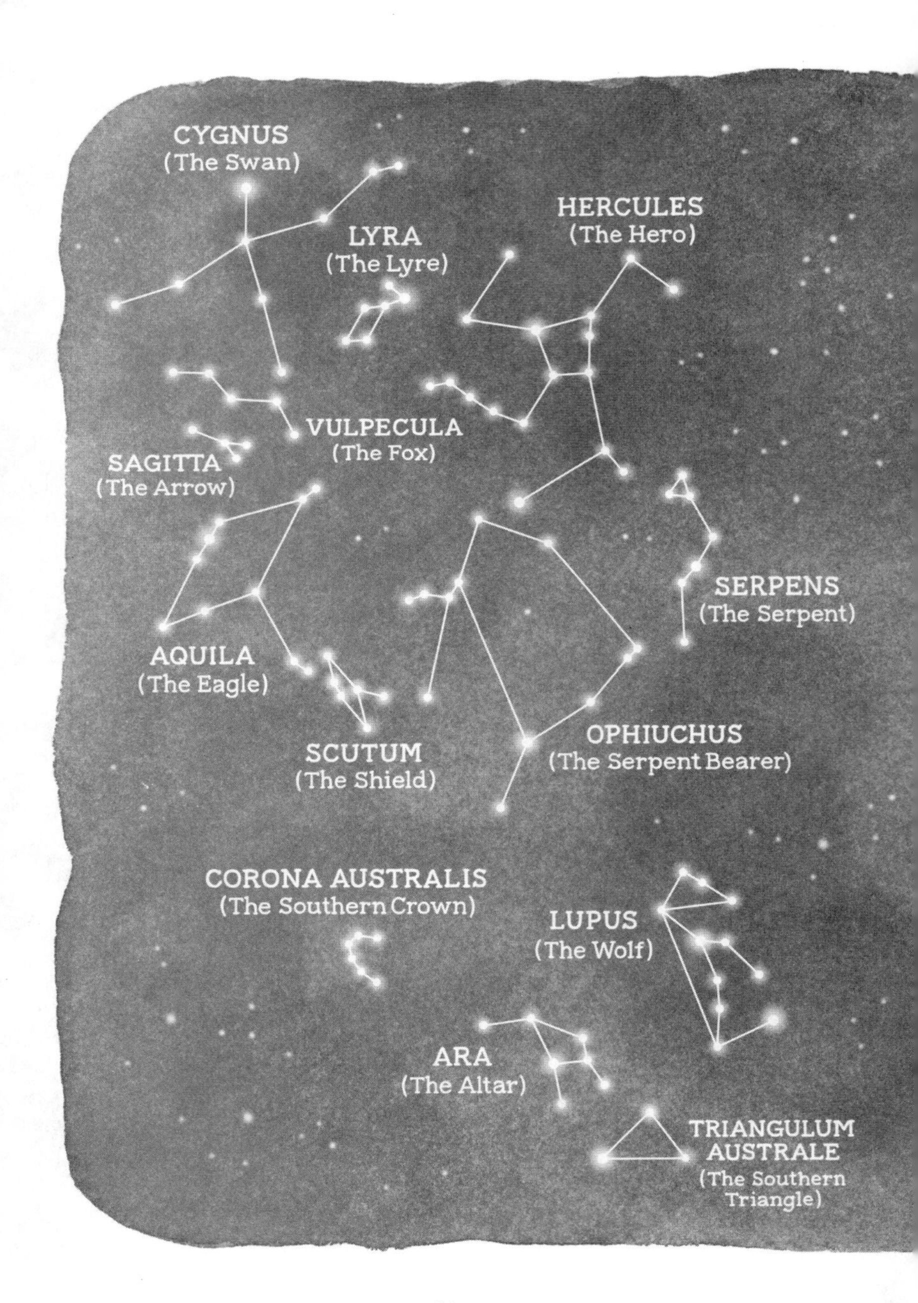
CYGNUS
(The Swan)
LYRA
(The Lyre)
HERCULES
(The Hero)
VULPECULA
(The Fox)
SAGITTA
(The Arrow)
SERPENS
(The Serpent)
AQUILA
(The Eagle)
OPHIUCHUS
(The Serpent Bearer)
SCUTUM
(The Shield)
CORONA AUSTRALIS
(The Southern Crown)
LUPUS
(The Wolf)
ARA
(The Altar)
TRIANGULUM
AUSTRALE
(The Southern
Triangle)

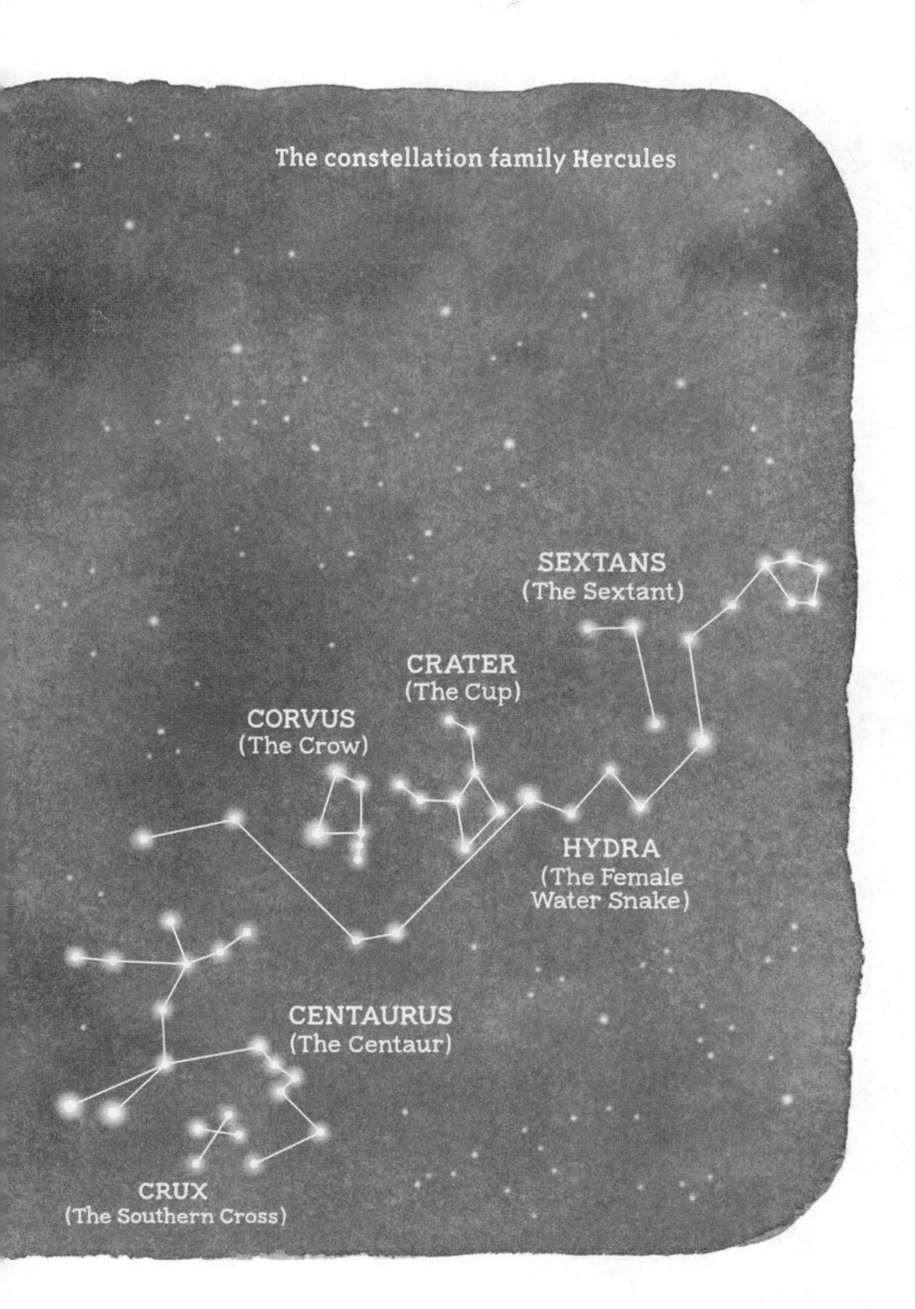
The constellation family Hercules
SEXTANS
(The Sextant)
CRATER
(The Cup)
CORVUS
(The Crow)
HYDRA
(The Female
Water Snake)
CENTAURUS
(The Centaur)
CRUX
(The Southern Cross)

If you are in the Southern Hemisphere, it's easier to spot Hydra, another constellation in the Hercules family. Hydra is the largest constellation in the night sky and looks like a water snake. Hydra had nine heads—Hercules defeated her in one of his many battles.

Another notable constellation in the Hercules family is the smallest one—Crux, or Southern Cross. People in the Southern Hemisphere use it for navigation, just as people in the Northern Hemisphere use the North Star. Five countries

(Brazil, Australia, New Zealand, Papua New Guinea, and Samoa) find this constellation so meaningful that it's on their flags. Different cultures tell different stories about the Southern Cross. The Kalapalo people of Brazil see Crux as a swarm of bees exploding from a beehive, while many Australian native people see it as part of a larger constellation that is an emu crossing the sky. An emu is a bird native to Australia. It can't fly, but it can run thirty miles per hour. The Tswana people of southern Africa see giraffes whose long necks reach into the sky to nudge the sun if it loses its way.

Flag of Papua New Guinea

The Orion family has the fewest constellations—only five. Although it's located in the Southern Hemisphere, it can be seen by stargazers in the Northern Hemisphere since it's so close to the equator. The constellation Orion shows a tall Greek hunter, wearing a belt made of three bright stars. He carries his shield and heavy club and wears a sword at his waist, while his two hunting dogs (Sirius is one of them) stay close by.

Two giant stars (one of which is Betelgeuse) make up Orion's shoulders, and another two form his knees. Orion is easy to recognize since it is so big and bright. The Aztecs called the stars in Orion's belt and sword the Fire Drill. Every fifty-two years they held their New Fire Ceremony.

Aztec New Fire Ceremony

This was to ensure the world wouldn't end. The ceremony occurred at the same time Orion's belt rose in the sky.

The constellation family Orion
ORION
(The Hunter)
LEPUS
(The Hare)

The next three families (Heavenly Waters, Bayer, and Lacaille) contain thirty-three constellations among them. All three of these families are located in the Southern Hemisphere,

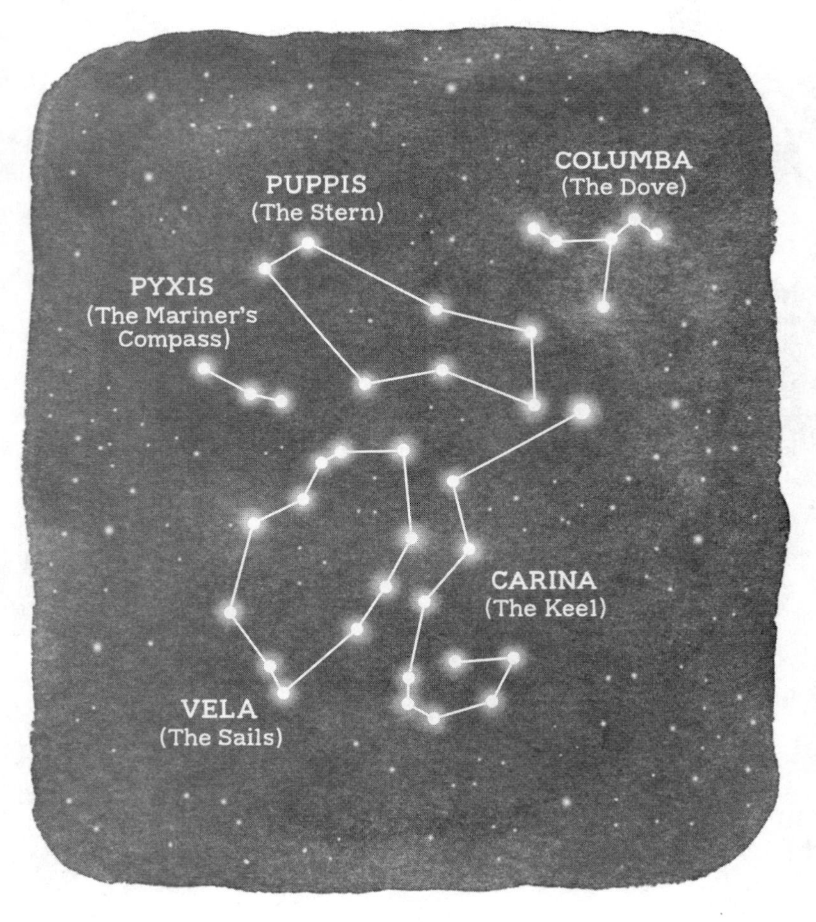

Heavenly Waters constellations in the Southern Hemisphere

except for a part of Heavenly Waters that is in the Northern Hemisphere. Their constellations aren't as well known as some others. The Heavenly Waters family contains nine constellations—all water-themed.

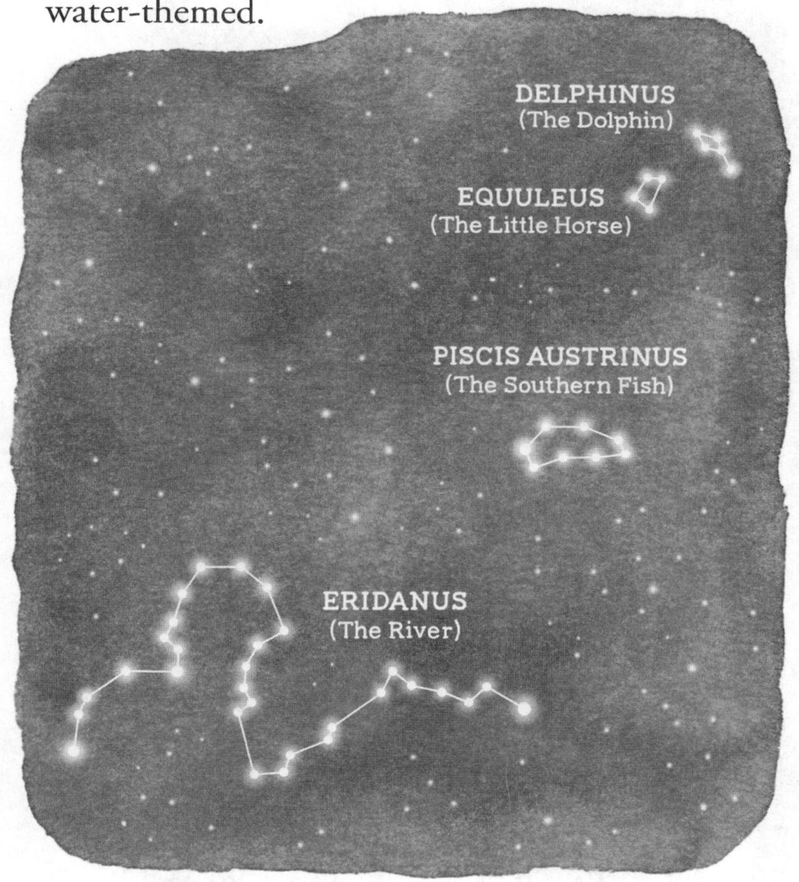

Heavenly Waters constellations in the Northern Hemisphere

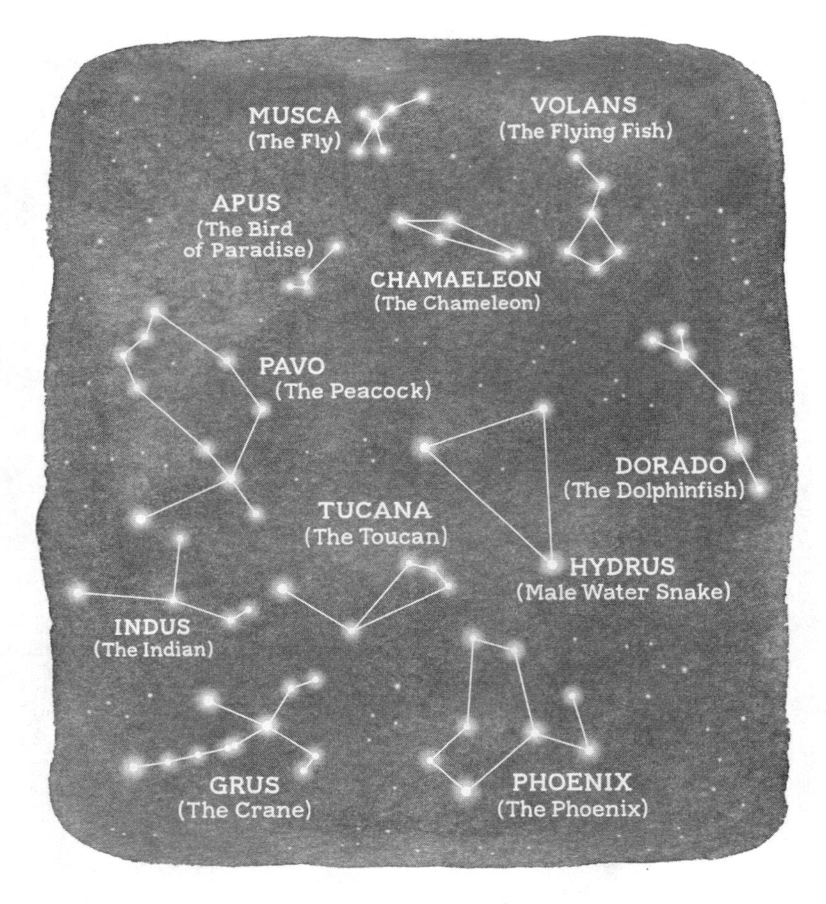

The constellation family Bayer

The Bayer family was named after the mapmaker Johann Bayer, and includes eleven constellations. And the Lacaille family was named after the astronomer and includes thirteen of the constellations he identified, named after tools like the clock, telescope, and microscope—all essential tools for making discoveries.

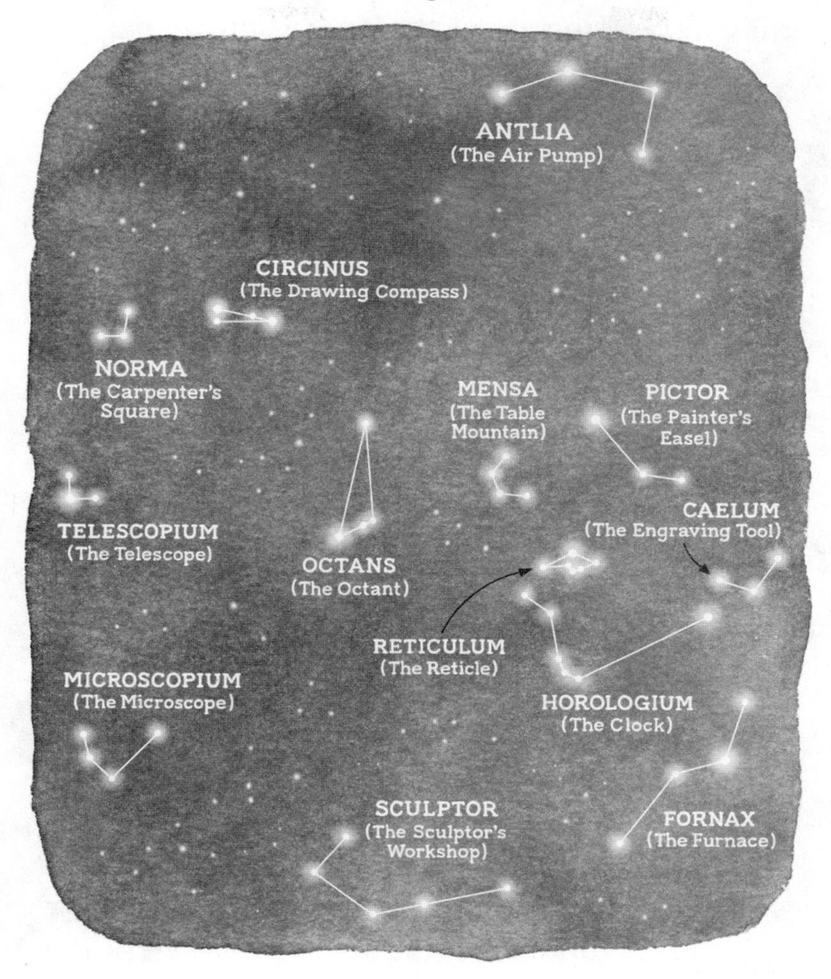

The constellation family Lacaille

That leaves one final constellation family—the Zodiac. The Zodiac is one of the most intriguing constellation families. Perhaps you know something about the Zodiac already.

CHAPTER 6
The Zodiac

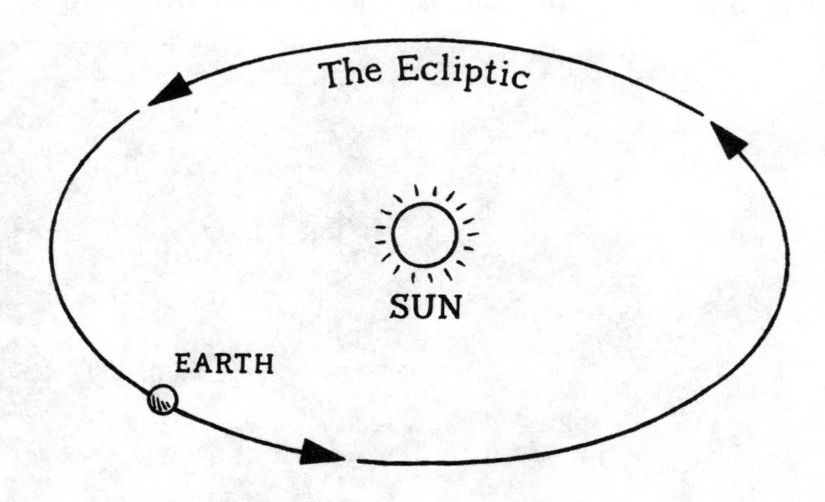

From Earth, the sun seems to follow a set path in the sky throughout the year, a path that changes slightly with the seasons. This narrow path is called the ecliptic. The twelve constellations in the Zodiac family are located in the ecliptic. (*Zodiac* is a Greek word that means *circle of animals.*)

They form a giant ring that wraps around Earth. Many of the constellations depict an animal-like creature.

Each of the Zodiac constellations represents a time of year that is about a month long.

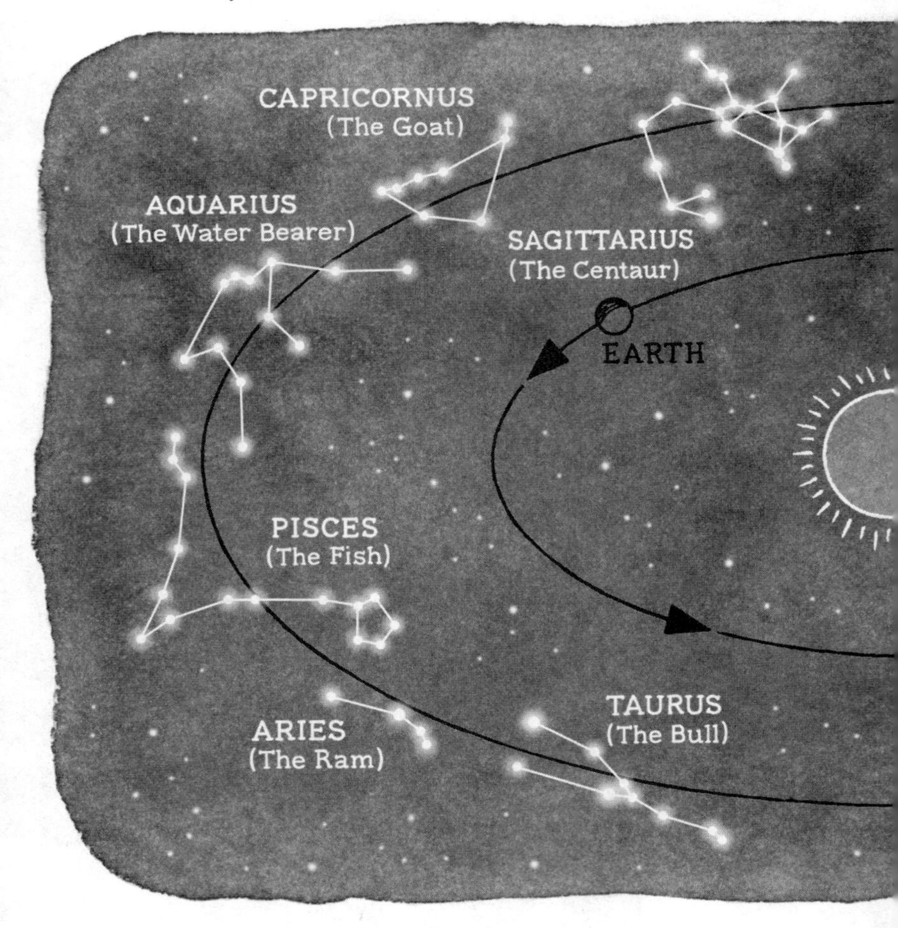

As Earth completes its yearlong orbit around the sun, a different Zodiac constellation becomes more visible each month. These constellations are visible to people in both the Northern and Southern Hemispheres.

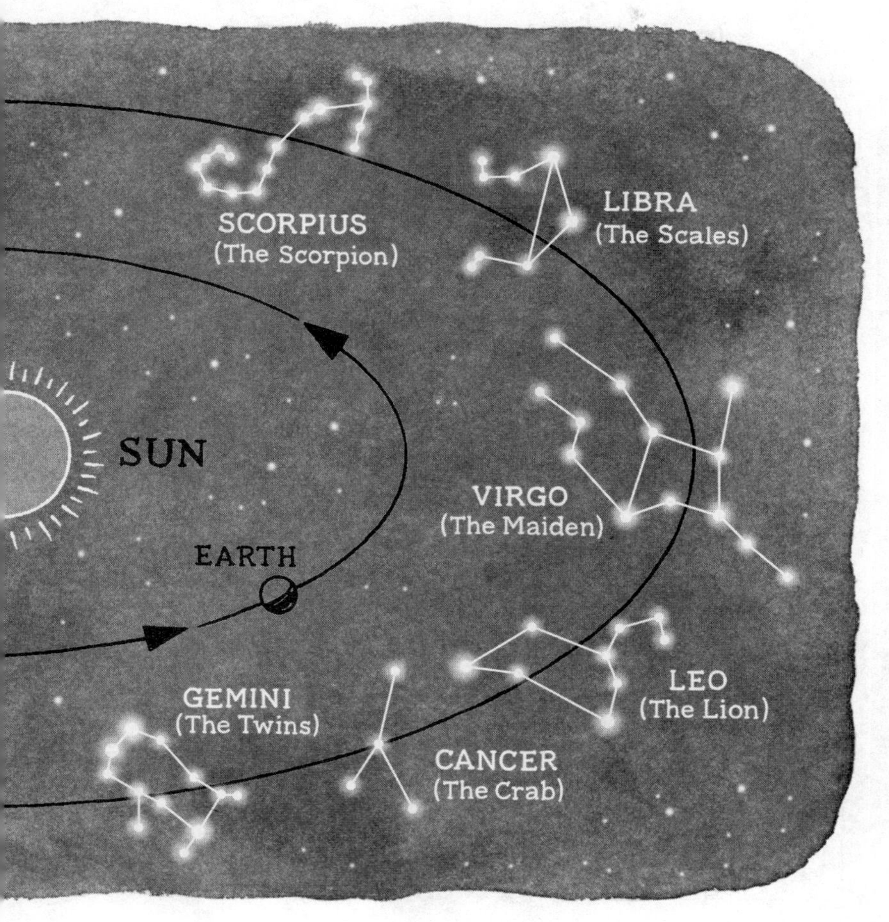

The Chinese Zodiac

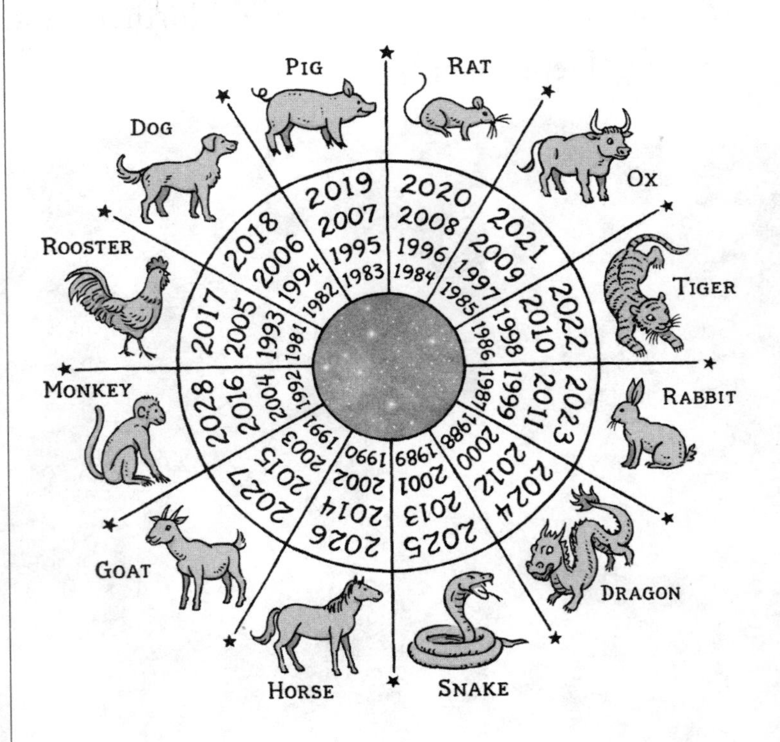

Perhaps you have heard of the Chinese zodiac. Is it the same as the zodiac just described? No. But there are similarities. Both zodiacs have twelve different symbols, mostly named after animals—but the signs of the Chinese zodiac correspond to years, not months. This is why in Chinese astrology it's more important what year you were born—for instance, the year of the goat or the year of the horse—than what month.

The Chinese zodiac comes from an ancient myth. In the sixth century BC, the Jade Emperor called all the animals to a race. The twelve animals that responded became the zodiac—each ranked in order of how fast they finished the race.

The first Zodiac constellation is called Aries, which is Latin for *ram*. The Aries constellation corresponds with the dates March 21 to April 19. Why is the first sign in March and not January?

Aries the ram

The spring equinox occurs around March 21 and is the sign for new beginnings. Like a new year. Also, our year—which begins in January—was devised by the ancient Romans at a later time in history. According to Greek legend, Aries the ram had wings made of sheep's wool that allowed him to fly. The brightest star of Aries is known as Hamal. That is not a Greek word. It's an Arabic word, from the phrase *ras al-hamal*, which means "head of the ram."

Taurus, the bull, is the next constellation. Within Taurus is a star cluster commonly called the Seven Sisters. Many cultures besides that of ancient Greece have stories about the Seven Sisters. They come from native Australians, the Hindus, and the Japanese. The Japanese call the Seven Sisters *Subaru*, which means "united." Subaru is also a name for a Japanese carmaker. The Subaru logo shows six of the stars.

One of the stories the Australian Aboriginals told was of seven sisters who shared a secret—they knew how to make fire. They carried long sticks with burning coals at their ends and used the fire to cook delicious yams.

One day, the sneaky Crow tricked the sisters into showing him how they made the fire. But it literally backfired. Crow as well as all his land burned up. The sisters were raised to the sky, where their burning sticks appear as stars.

The Gemini constellation is also known as the Twins (in Latin, *Gemini* means *twins*). It contains two very bright stars, named Castor and Pollux. The Greeks believed Castor and Pollux were twin brothers—one was immortal (he would live forever) and one was mortal (he would die).

Ancient Greeks thought Cancer looked like a crab that used its claws to grab Hercules as he battled Hydra, the water snake. Hercules crushed Cancer, but the goddess Hera—who was Hercules's enemy—rewarded the crab by letting it stay in the sky.

Leo, also called the Nemean Lion, fought Hercules but was defeated. Hercules took Leo's impenetrable hide and used it as a cape. This constellation is known as the Lion by the Syrians and also the Turks.

The sixth constellation is Virgo, the maiden. She lies down in the sky holding an ear of corn, representing the harvest season for the Greeks. Virgo is the second largest constellation in the sky after Hydra, which is in the Hercules family.

Libra does not look like a living creature. Instead, it looks like a set of scales and represents balance. Libra is most visible in October, a time of year when days and nights are roughly of equal length. It is believed that's how it got its name. Libra contains the star Methuselah, which is the oldest star in the night sky. Methuselah is thought to be more than thirteen billion years old—nearly as old as the universe! It was named after the oldest person in the Bible—a man who lived 969 years!

The Scorpius (or Scorpio) constellation definitely resembles what it's named after—a stinging scorpion! Scorpio and Orion are not visible in the sky at the same time. That's because they are enemies. According to a Greek myth, Scorpio killed Orion!

Centaur

Sagittarius and Capricornus (or Capricorn) come next and are similar because they are both hybrids (a combination of two different creatures). Sagittarius is a centaur. In Greek mythology, a centaur was half man, half horse. Sagittarius holds a bow and arrow. Legend has it that Hercules shot Sagittarius by accident with a poisoned arrow. Although Sagittarius was injured badly, he didn't die because he was immortal. Capricorn is half goat, half fish. Two ancient civilizations—the Babylonians and Sumerians—both called Capricorn the goat-fish.

Aquarius is known as the water bearer. (The Latin word for water is *aqua*.) Aquarius pours his water into the mouth of the constellation Piscis Austrinus (in the Heavenly Waters family). In some parts of the world, Aquarius is most visible in the rainy season, which may explain why the constellation is linked with water.

The last constellation of the Zodiac family is Pisces. There are a few different stories about Pisces. The Babylonians believed it was two fish joined together by a cord. The Romans believed one of the fish was the goddess of beauty, named Venus, and the other was her son, Cupid. The Greeks believed this constellation was the sign that spring was about to begin.

Have you ever had someone ask you what sign you were born under? The answer would be the name of one of the Zodiac constellations. For instance, if your birthday is between April 20 and May 20, then you're born under the Taurus sign. Some believe that your sign tells a lot about what you're like and influences what may happen in your life. This belief goes back thousands of years.

The study of how stars influence the future is called astrology. The name *astrology* sounds very much like *astronomy*. But astronomy is a science. Astrology is not. Still, many people today read their daily horoscope in a newspaper or online for fun. *Horoscope* comes from the Greek words meaning "hour" and "observer." A horoscope is a prediction of the future. It is based both on the day you were born and where the stars and planets are positioned on the current day.

Certain personality traits are associated with each sign of the zodiac. Do you think that your sign describes you? Or is it way off base?

Are They Correct for You?

♈ Aries (3/21–4/19): bold, ambitious

♉ Taurus (4/20–5/20): stubborn, loyal

♊ Gemini (5/21–6/21): lively, open-minded

♋ Cancer (6/22–7/22): thoughtful, emotional

♌ Leo (7/23–8/22): enthusiastic, proud

♍ Virgo (8/23–9/22): logical, critical

♎ Libra (9/23–10/23): friendly, fair

♏ Scorpio (10/24–11/21): secretive, passionate

♐ Sagittarius (11/22–12/21): generous, restless

♑ Capricorn (12/22–1/19): resourceful, cautious

♒ Aquarius (1/20–2/18): independent, rebellious

♓ Pisces (2/19–3/20): imaginative, sensitive

CHAPTER 7
Stargazing

It is believed that astronomy is one of the oldest sciences. A circle of stones that might have been an observatory was built in present-day Turkey in 9000 BC. That's over eleven thousand years ago. (An observatory is a location or a structure specially designed to observe what's happening in the sky.) Huge monuments of stone circles can be found in parts of the world like Egypt, Brazil, and Australia. Other ancient circles of stones have been found in Russia, Spain, and England. All of them may have been used as observatories.

The circular structure of stones in southern England is the most famous—it's called Stonehenge. The stones are huge—thirteen feet tall—and weigh about twenty-five tons. We are not sure who built Stonehenge, but scientists believe it may have been used to monitor the movement of the sun and moon. Stonehenge is thought to be more than four thousand years old.

Stonehenge

In modern times, there are two different kinds of observatories—those on Earth and those located in the sky. There are hundreds of observatories located on all seven continents, even Antarctica! They are built on areas of high elevation (like atop mountains) and away from cities, because it's easier to see many stars where the air isn't filled with pollution or electric light. Astronomers study and work at these observatories, but tourists are allowed to visit some of them, too.

One of the largest modern-day observatories is at Mauna Kea, located on the island of Hawaii. Mauna Kea is an inactive volcano whose peak is fourteen thousand feet above sea level. The skies are clear 90 percent of the time at Mauna Kea. Twelve organizations have set up operations on Mauna Kea, and it has thirteen different facilities with high-powered telescopes.

Researchers at Mauna Kea have made many

Mauna Kea Observatory

discoveries over the years. In 2005, scientists studying the movement of the stars in the Andromeda galaxy realized the galaxy was three times as large as they had originally thought. Also in 2005, scientists discovered that Mars was releasing the gas methane. This is important because it means that some sort of life could exist on Mars!

Paranal Observatory

The South American country of Chile has more than a dozen observatories. One of the most famous is Paranal Observatory, located deep in the Atacama Desert—one of the driest places in the world. Paranal is home to the Very Large Telescope (VLT), which actually is made of four separate telescopes. The VLT was one of the first telescopes to register the dimming of the star Betelgeuse.

What about observatories *not* located on Earth? These are craft with no crew aboard that travel in space and send information back to Earth. The most well-known is the Hubble Space Telescope, named after American astronomer Edwin Hubble.

The Hubble was launched in 1990 and has been orbiting Earth ever since. It completes a full orbit of Earth every ninety-five minutes—traveling five miles every second! The Hubble has sent back over a million pieces of scientific data, including information that has helped astronomers confirm the age of the universe and learn about seasons on other planets.

Edwin Hubble (1889–1953)

Edwin Hubble was born in 1889 in Missouri and grew up in Chicago. As a young boy, he was fascinated with the sky. Instead of having a birthday party when he turned eight, he chose to stay up all night with his grandfather's telescope. Looking back at his youth, he once said, "I knew that even if I were second- or third-rate, it was astronomy that mattered." Well, in fact, he became a first-rate astronomer. Hubble became known for many important findings, such as proving that there were more galaxies in the universe than just the Milky Way. He also showed how some galaxies are constantly moving away from one another. His research helped form the big bang theory. The Hubble Space Telescope was named in his honor.

The Hubble also discovered two of the five moons orbiting the dwarf planet Pluto. In addition to these scientific data, the Hubble has also sent back the first gorgeous images of deep space ever seen.

You don't need to travel to an observatory or own expensive scientific equipment to stargaze. The farther away you are from cities, the easier it is to see the night sky. Sometimes all it takes is going into your backyard. You can look up and spot Venus or the Big Dipper with just your eyes. Other times an inexpensive star wheel will help. A star wheel is a paper map of the stars. Stargazers rotate the wheel to select the current date and time. Voilà! The star wheel shows what the sky looks like at that point in time.

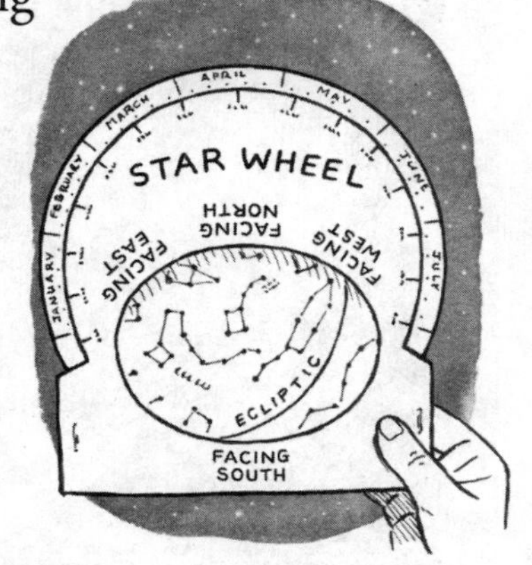

Is It a Star?

The Aurora Borealis (Northern Lights) occurs close to the North Pole, and the Aurora Australis (Southern Lights) occurs close to the South Pole. People travel far to catch a glimpse of the colorful lights that stretch in bands across the night sky. Most of the time the aurora glows white or green. But sometimes it is a spectacular shade of pink, purple, and red. Some people think these are colorful stars,

but they are not stars at all. Why does the sky glow these colors? Tiny particles stream from the sun and hit Earth's atmosphere. These particles give their energy to molecules in the atmosphere. But those molecules can't hold that energy, and convert it into light particles instead. The result is a beautiful display of light. The Vikings thought the Northern Lights were the shining weapons of warriors. One legend of the Alaskan Inuit said they were the souls of animals like salmon and other fish.

View of Aurora Borealis in Norway

There are also hundreds of apps that help stargazers. Some of them allow users to hold their device's camera up to the sky and the app identifies what is seen. And, of course, planetariums are great places to learn more about the stars. Planetariums are huge dome-shaped theaters. A projector shines images on the dome while you listen to the explanation of what you are seeing. There are hundreds of planetariums in the United States alone. Many are located at science museums or universities. Find out if there is a planetarium near you.

Hayden Planetarium in New York City

Stargazing is an ancient pastime still beloved by humans today. New tools are constantly being developed to study the stars. The late British scientist Stephen Hawking encouraged stargazing, saying, "Remember to look up at the stars and not down at your feet. Try to make sense of what you see and wonder about what makes the universe exist. Be curious."

The starry sky is a magical place. Humans have been looking upward since the beginning of humankind. And everything we can see (and can't see) in the sky has been around much longer than any of us. We will continue to look up to the sky with amazement and wonder—and always wish upon a star.

Timeline of the Constellations

mya = million years ago

c. 10 mya	Betelgeuse the star begins to form
c. 2500 BC	Stonehenge is built in England
c. 350 BC	Chinese astronomer Shi Shen compiles one of the first known star catalogs
c. AD 150	Ptolemy identifies more than a thousand stars that make up forty-eight constellations in the *Almagest*
1603	German lawyer and astronomer Johann Bayer creates *Uranometria,* a star atlas
1919	The International Astronomical Union (IAU) is formed
1922	The IAU comes up with eighty-eight official constellation names
1975	Donald Menzel groups the eighty-eight constellations into eight separate families
1990	The Hubble Space Telescope is launched into orbit
2005	Researchers at Mauna Kea Observatory realize the Andromeda galaxy is three times as large as they originally thought
2008	A teenager in New York State is the youngest person ever to spot a supernova
2019	Astronomers notice that Betelgeuse has dimmed considerably

Timeline of the Universe

bya = billion years ago

c. 13.8 bya	The universe begins to expand from a single point
c. 4.6 bya	Earth begins life as a small rocky object
c. 3.8 bya	The first life appears on Earth
c. 2.3 mya	The first humanlike creatures appear on Earth
c. 200,000 years ago	The first "modern" humans appear in Africa
c. AD 100	Ptolemy is born in ancient Egypt
1543	Copernicus shares his belief of a sun-centered solar system
1608	Hans Lippershey invents the first telescope
1609	Galileo's research supports Copernicus's sun-centered solar system
1781	William Herschel discovers Uranus
1846	Neptune is discovered
1889	Edwin Hubble is born in Missouri
2005	Scientists at Mauna Kea discover that Mars is releasing methane
2006	Pluto is demoted to dwarf planet

Bibliography

***Books for young readers**

*Aguilar, David A. ***Space Encyclopedia: A Tour of Our Solar System and Beyond.*** Washington, DC: National Geographic Kids, 2013.

Barker, Sarah. ***50 Things to See in the Sky.*** Hudson, NY: Princeton Architectural Press, 2019.

Davies, Alison. ***Written in the Stars: Constellations, Facts, and Folklore.*** London: Quadrille, 2018.

*de Mullenheim, Sophie. ***Earth & Sky! Tell Me Books.*** Hauppauge, NY: BES Publishing, 2018.

*Driscoll, Michael. ***A Child's Introduction to the Night Sky.*** New York: Black Dog & Leventhal, 2004.

Petersen, Carolyn Collins. ***Astronomy 101: From the Sun and Moon to Wormholes and Warp Drive, Key Theories, Discoveries, and Facts About the Universe.*** Avon, MA: F&W Media, 2013.

*Sabol, Stephanie. ***Where Is Our Solar System?*** New York: Penguin Workshop, 2018.

*Stott, Carole. ***I Wonder Why Stars Twinkle and Other Questions About Space.*** New York: Kingfisher, 2011.

Websites

www.iau.org

www.nasa.gov

AQUARIUS
SAGITTARIUS
PEGASUS
CAPRICORNUS
AQUILA
ANDRO
CASSI
CYGNUS
CEPHEUS
HERCULES
URSA MINOR
DRACO
OPHIUCHUS
SCORPIUS
LIBRA
VIRGO
CENTAURUS

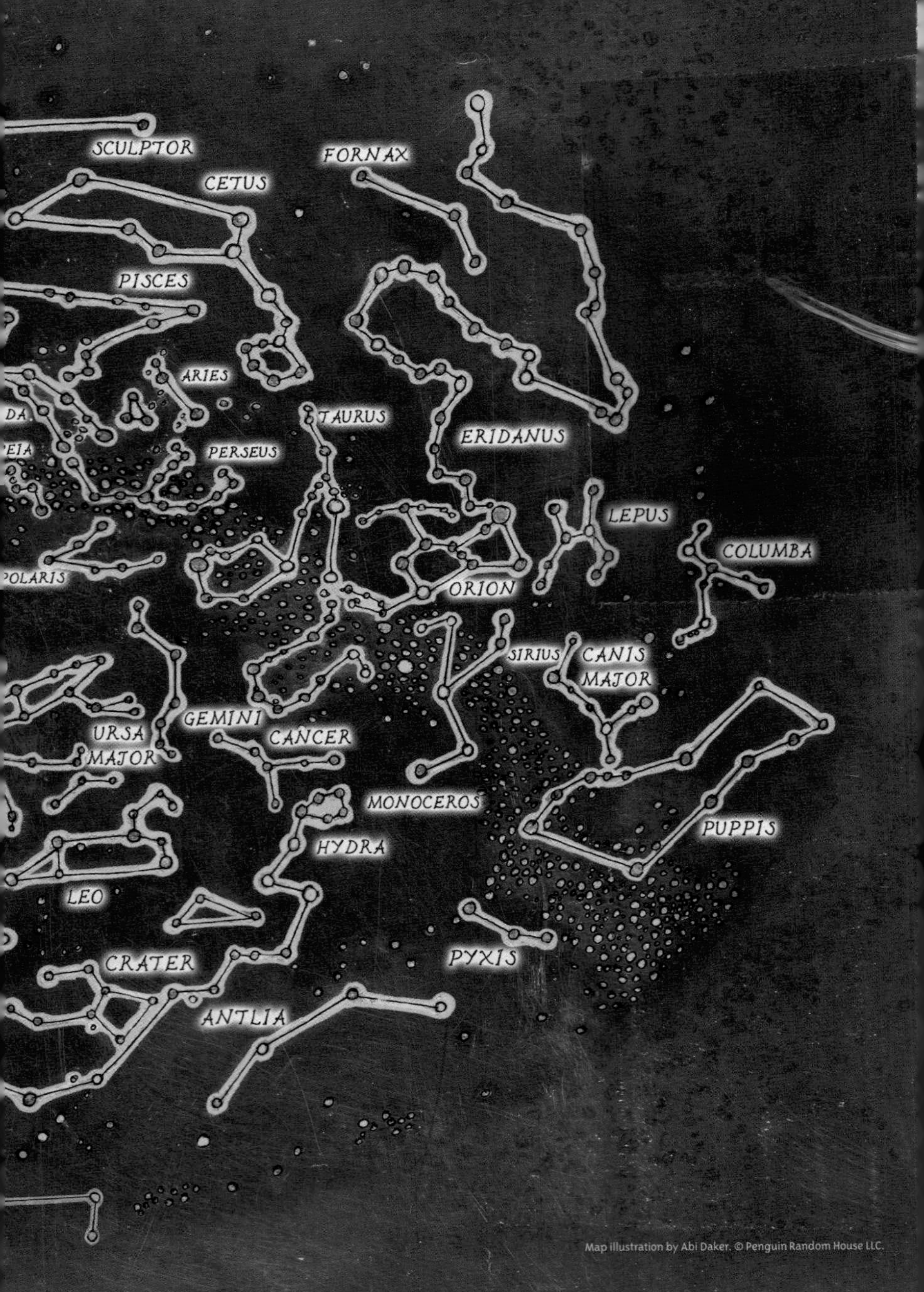
SCULPTOR
FORNAX
CETUS
PISCES
ARIES
TAURUS
ERIDANUS
PERSEUS
LEPUS
COLUMBA
POLARIS
ORION
SIRIUS
CANIS
MAJOR
GEMINI
URSA
MAJOR
CANCER
MONOCEROS
HYDRA
PUPPIS
LEO
CRATER
PYXIS
ANTLIA